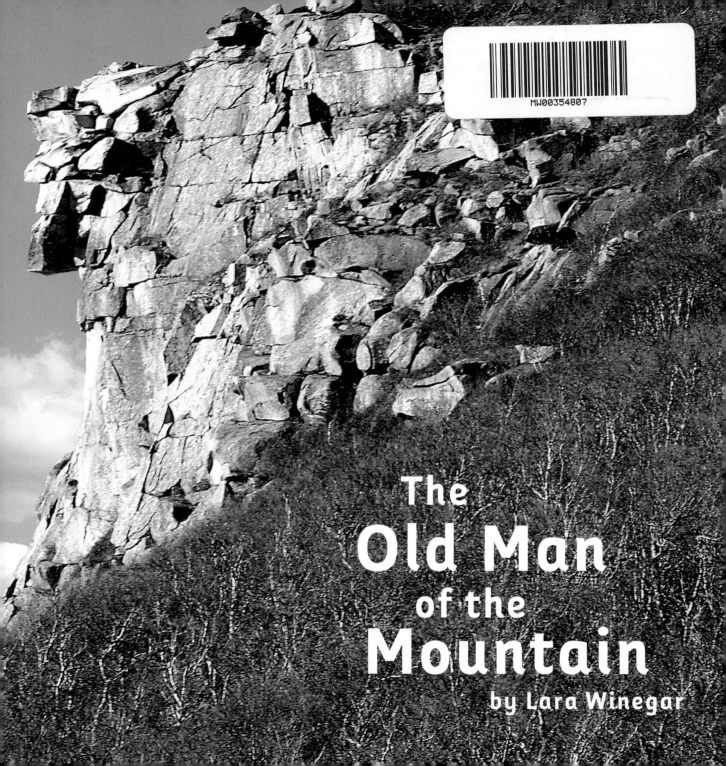

The Old Man of the Mountain

by Lara Winegar

Once there was a famous rock.

The rock was big.

Old Man of
the Mountain

New
Hampshire

It was part of a mountain. The
mountain is in New Hampshire.

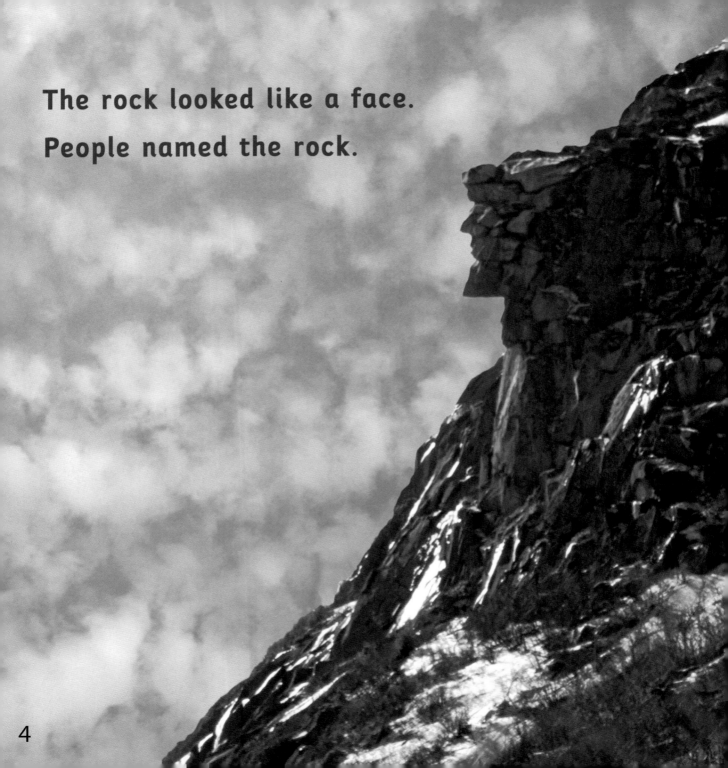

The rock looked like a face.
People named the rock.

e

se

in

wrinkle

They called it the Old Man of
the Mountain.

5

People took pictures of the rock.

The rock had cracks. The cracks looked like wrinkles!

rock

ice

Water got in the cracks. The water froze into ice.

The ice made the cracks bigger.
This is called weathering.

People tried to fix the cracks.

But the rock cracked more.

10

The face
was here.

Pieces of the rock fell from the
mountain. Gravity pulled the pieces
to the ground.

This happened in 2003. But you can still see the Old Man today on coins and gifts.